ANTS

ants
ants

ants

JOACHIM OFFENBERG

ar

ants

E
LEC
I
N

ANTS

Layout and cover: Camilla Jørgensen, Trefold
Cover photograph: Poul Ib Henriksen
Publishing editor: Søren Mogensen Larsen
Translated from the Danish by Heidi Flegal

This book is typeset in Dante and Gotham and printed on Munken Premium Cream

Printed by Narayana Press, Gylling, Denmark
Printed in Denmark 2019

ISBN 978 87 7184 813 7
Part of the Reflections series from Aarhus University Press

Published with the financial support of Aarhus University Research Foundation

AARHUS UNIVERSITY

Aarhus University Press
Finlandsgade 29
DK-8200 Aarhus N
Denmark
www.unipress.dk

International distributors:
Oxbow Books Ltd.
www.oxbowbooks.com

ISD
www.isdistribution.com

PEER REVIEWED

MIX Paper FSC FSC® C010651

CONTENTS

WORLD CHAMPIONS IN EVOLUTION

IF YOU WERE A MARTIAN

Imagine you were a little green Martian gazing down on Earth. What would you think of humans, these odd creatures making such a huge impact on the planet? And what would you think of ants? Perhaps the ant would garner more attention than might seem reasonable to us. After all, we *Homo sapiens* are used to perceiving ourselves as masters of the globe. But to a stranger looking down from space, our dominance is not so obvious. Ants, you see, are a fair match for humanity, not only in numbers but also in terms of weight and evolutionary sophistication.

Determining the precise number of humans and ants is no easy task, but even a rough comparison shows that ants win, hands down. The world is home to some 10–100 billion *billion* ants, so even if we miscalculate their number by a factor of 100, or by a billion individuals, we are still vastly outnumbered. Looking instead at total weight or 'biomass', the ants' lead dwindles, yet even the most conservative scientists estimate that the planet's formican biomass exceeds its human biomass. Others estimate that ants outweigh humans by up to 100 times. Again, ants apparently take the gold. In our defence, the global ant community consists of about 12,000 described species, whereas modern humans all belong to one species. At any rate, ants and humans are both indisputable heavyweights if we measure biological success in tonnage, boasting 900–9,000 and 420 million tonnes of biomass, respectively. By comparison the world's largest living animal – the blue whale – isn't even close to the biomass bronze, weighing in at a measly 0.5 million tonnes.

ANTS ARE LIKE US

Size may matter, but not as much as you might think. Imagine Martians zooming in for a closer look at the lives of our two races. Would they acknowledge humans as superior to ants? Probably not. They would find that humans are not the only life form with highly advanced social structures. Ants can also subdue, control and manipulate other organisms, and they too use medicinal substances. They can even adapt to deal with an unpredictable future through rationing, food storage and cannibalism. Also, the social evolution of ants is so highly refined that they practice complex – albeit nasty – manoeuvres such as *coups d'état* and slave raids.

Humanity is not the epitome of an evolutionary process. In many ways we are simply a delayed parallel product that would hardly impress a curious Martian. Humans are not alone in using agriculture and animal husbandry. Nor are we alone in collecting and transporting energy or reaching our goals by means of clear communication, efficient collaboration and the use of tools. Ants do all this too, and probably have done for much longer than us – since ants have been around for more than 100 million years, whereas *Homo sapiens* arrived on the scene just 200,000 years ago. In other words, when we study ants we are not simply studying insects. We are observing a precursor to our own success.

MEET YOUR NEW FRIENDS

Down through the ages, humankind has been fascinated by the ant's life and habits. I've been told that virtually every language has a word for 'ant', though other animals often go unnamed. In many cultures sayings about ants abound, as in the Bible (Prov. 6:6): "Go to the ant, thou sluggard; consider her ways, and be wise". We know ants because they are everywhere, making their mark on the world, for like us, they can change their environment through their industriousness and strength in numbers. They conquer our homes with infuriating efficiency in their never-ending quest for resources. They burrow under our patios and paving stones so skilfully that we spend fortunes on pesticides, vainly attempting to keep them at bay. They chase us away in coordinated attacks if we spread our picnic blankets too close to their colonies. No wonder most people know ants well, and detest them. The funny thing is, the characteristics that make us hate them most – their appetite for conquest and their efficient exploitation of their surroundings – closely mirror distinctive human traits.

This little book takes you on a quick guided tour into the great wide world of ants, showing just how much it resembles our own world. This journey is not merely fascinating in itself. It also explains why ants and humans are among the most successful beings on Earth, and by zooming in on our similarities we reveal essential ingredients in the recipe for biological success.

If these were reflections on management and leadership, I could have interviewed prominent figures in the Danish business community, captains in some of our small country's largest enterprises – Maersk, Lego and Vestas – hoping they would divulge some of the secrets behind their commercial success. But the prominent figures in this book are actors in a very different arena, on a very different scale. Meet the weaver ant, the honeypot ant, the wood ant, the leafcutter ant, the stinger ant, the headcutter ant and the black garden ant. You will learn how they handle critical tasks just as ingeniously as humans do. And that's not all. Comparisons between our two races show a huge potential for mutually beneficial collaboration. In short, ants are actually your friends ... You just haven't been properly introduced yet.

THE BOONS OF SOCIAL LIVING

1 + 1 = 3

Have you ever been alone in a stalled car that needs a push to get the engine started? If so, you will know that, alone, you face an impossible task. Now consider how rarely we think about the countless other things we do that would be impossible without social behaviour. To us, living around other members of our species is a way of life so deeply ingrained we find it hard to imagine anything else. But most species are *not* social animals. This seems rather odd, given that 'socialism' has such enormous prospective benefits. It is striking, for instance, that the two large groups of insects topping the biomass list – ants and termites – are the insect orders where all species are social. The advantage of social living is collaboration, which can create synergy. Two individuals can achieve *more* than twice as much as one individual on its own.

WORLD-CLASS ARCHITECTURE

From the Sydney Opera House to the skyscrapers of Tokyo and the Eiffel Tower in Paris, humankind's ability to build extraordinary structures is amazing. Even small structures like a humble, terraced house in a suburb require a host of high-level specialists, from planning through to completion. Ants have also mastered the art of construction, going much further than the little ant hills every child knows so well. Leafcutter ants and weaver ants are just two of the many species that have developed building practices of astonishing size and complexity. One look into their homes makes it clear that cooperative housebuilding is an important part of their biological success.

THE SUBTERRANEAN METROPOLIS

Leafcutter ants live in colonies of up to 20 million inhabitants, corresponding to the largest human conurbations. Populations of this size need lots of space, and a handful of myrmecologists – ant scientists – who were tired of pencil-pushing and sterile laboratories decided to find out just how large such a colony was. They stirred up a thin mixture of liquid concrete and poured it into a leafcutter colony's entrance holes, letting it flow down into the passages and chambers. After a few days of arm-wrenching work, they brought in a mixer truck with a whole load of thin concrete, which soon disappeared, so they ordered another ... and poured until the concrete began to spread out above ground. Glad to rest their arms while the concrete hardened, they knew they would be spending days digging. Using excavators, shovels and trowels they set to work removing the soil around the concrete cast – finding that the colony stretched several stories down. Working for days from a growing lattice of scaffolding, the group gradually uncovered the colony's shape, revealing an astounding underground metropolis that looked more like science fiction than earthbound fact.

RUBBISH DUMP OR RETIREMENT HOME?

Leafcutter cities can span an area of 50 square metres and run to a depth of 6 metres. To construct such a subterranean habitat the ant colony must remove some 40 tonnes of soil, creating a wide variety of chambers with various functions. Some are brood chambers, others are fungus gardens. Yet others, at the outskirts of the complex, are rubbish dumps tended by waste-management workers.

Because these chambers contain spent compost, dead ants and other detritus, they harbour rot and decay, meaning a risk of infection for the ants working there. Waste workers are barred from the centre of the metropolis, moving only up to the transition area between their own waste chamber and the residential sectors. Here, other ants hand them the rubbish, which they dutifully carry the last stretch to the dump. In a society geared to promote maximum efficiency, the most expendable individuals are appointed to do high-risk jobs like handling waste. Among ants, the most expendable workers are those destined to die soon anyway. Rather than a retirement home or hospice, the ant colony demotes the sick and elderly to work in the rubbish dump – so despite many similarities, there are also tremendous differences between humans and ants.

LIFE IN THE PENTHOUSE

Then there is the weaver ant, which lives in colonies whose collaborative efforts afford them a lofty life, high above the dirt-ridden, dangerous world of their predominantly ground-living cousins. They have moved into the trees and overcome the biggest problem facing most tree-living ant species: finding appropriate nesting sites. Weaver ants use leaves as their nesting material, and because they live in the tropics, where trees don't lose their leaves all at once, the supply is continuous.

But before weaver ants could move from holes in the ground to airy penthouses, their collaboration had to make a quantum leap. The leaves on their tree must somehow be drawn together and fixed into a sturdy spherical structure. Given that workers of this species measure just 4–5 mm, this takes a major feat of engineering.

The solution is simple but ingenious. If a worker is unable to reach a leaf that the colony has selected as desirable, for reasons known only to the ants themselves (a matter that continues to baffle myrmecologists), the next worker to arrive will crawl over the first one. Then the two ants will hook themselves together, one ant's jaws gripping the other's waist. They continue this behaviour, forming a chain that enables them to reach their goal: the coveted leaf. Weaver ants can make chains over 20 centimetres long. To pull the leaves together, they build a slightly shorter chain next to the first one. Hundreds of ants work together, constructing a series of ever-shorter chains and finally bringing the leaves together. Then comes the next challenge: keeping the leaves together.

GRUBBY GLUE GUNS

First, worker ants tasked with holding the leaves together lock themselves into position like so many tiny safety pins. Their feet cling to one leaf edge like Velcro, while their pincer-like jaws grip the opposing edge. There they sit, for hours on end, while the worker ants tasked with fixing the leaves in place scurry away to pick up their junior assistants: ant grubs. Unlike adult workers, ants in the larval stage can spin silk. Held firmly in the jaws of a worker, a larva is moved back and forth, from edge to leafy edge, spinning a silken thread that it dutifully fastens each time the worker presses its head against the leaf. How the larva ever learned to do this remains a mystery.

So here we have ant larvae serving as glue guns – an extremely rare example of insects using tools. Such practices, seldom seen in the animal kingdom, normally occur among birds, primates and other higher-order animals.

Weaver-ant engineering results in nests the size of beach balls, and unlike other ant species they never have trouble finding nesting sites. Compared to solitary insects, weaver ants have competitive advantages that are achievable only with sophisticated cooperation among several hundred individuals.

EXTREME NICHE SPECIALISATION

Ants have no vocational guidance counsellors, and young ants are not encouraged to follow their dreams when choosing a career. Ants are born to do certain tasks, much as our feudal societies once believed humans were born to fulfil certain roles. In the ant world specialisation equals efficiency, and individual workers can function as specific organs for the entire colony, quite literally. Whereas human societies are based on specialised *behaviours* – one person works as a police officer, another works as a baker – ant societies have taken this to extremes. They have developed various castes of individuals with a specialised *physique* that makes them well-suited to perform very particular jobs.

Among Danish ant species we only see differences in worker size. Further south, certain species have special soldier ants with massive, muscular heads capable of delivering a powerful bite when defending the colony's resources or attacking prey. Some species have just a single worker for one specific job – for instance a worker with an extremely specialised head shape that fits like a stopper into the entrance hole of the colony's hollow-branch nest. When this ant, one of the doorkeeper caste, sticks its head into the hole from the inside, the entrance to the colony is plugged tight so that enemies, and rising tidal waters from the surrounding mangrove swamp, cannot enter.

Ants are experts in extreme specialisation and niche occupations – a key factor in their ability to conquer very diverse habitats. Meanwhile, in the modern human world things seem to be moving in the opposite direction. Often we hear that each person must be versatile, flexible, adaptable, although during evolution we too used the division of labour to make great progress.

SEX FOR THE CHOSEN FEW

Systems to control reproduction are also a shared feature in human and ant societies. Many of our cultures have forbidden incest since the dawn of time, and in the modern health-care system prenatal diagnostics and quality-checking of semen and ova have become routine procedures. Here, too, ants have gone much further, applying a strategy which, while wholly unacceptable to humans, has proved extremely efficient for ants' geographical expansion. Actually, it is yet another example of just how radical formican specialisation is. The division of labour covers not only practical daily tasks. It extends to the rigorous control of reproduction itself. Only selected ants are permitted to breed, while the lower castes are relegated to helping them do so by tending the nest and brood. This extreme reproductive specialisation is based on a principle called *eusociality*.

Eusociality in its purest form is found in colonies where only one female is allowed to breed, as in the case of ants. The queen, the very core of the ant community, is the only one that can lay eggs and produce new individuals. If she dies the entire colony dies, since in most ant species the sterile workers have lost all ability to breed. Put simply, the queen is the ant colony's ovaries.

It is a huge advantage that, being its most vital organ, she can stay hidden deep inside the nest, safe and sound. The sterile workers, on the other hand, can be sacrificed without compromising the colony's survival, so they handle all the dangerous work outdoors, where enemies abound.

THE ROYAL SPERM BANK

In the sexual-reproduction game the prospects for the vast majority of female ants are bleak to non-existent. Even so, their fate as wilted wallflowers seems bright compared to the life of male ants – which are reduced to the role of flying sperm cells.

A male ant is nurtured to adulthood by the workers in his colony, but as soon as he is sexually mature he is sent off on a mating flight with all the new virgin queens from the area. Large swarms of these winged ants rise up almost simultaneously and gather in the twilight, and during the mating flight each male does his best to woo and mate with his one and only. And he gets only a single go. Once he has mounted and mated with a queen, he dies. The male ant is a good example of what life is basically all about for males: procreation. A male ant has only one shot at success, so he has to make it count if he is to pass his colony's genes on to the next generation. He is effectively the colony's testicles and penis, just as the queen is the colony's ovaries. So sex is good, and indeed it is the sole purpose of his existence. But for ants, more sex does not mean better sex, and once is enough if they get the job done right.

The trick is that the queen's body has a built-in sperm bank, a special gland where the semen can be stored throughout her lifetime, up to 30 years. Here the sperm cells are carefully sustained in a sort of suspended animation until being released and allowed to fertilise an egg. In some species the queen mates with and retains sperm from several males. Other species have faithful queens that only receive sperm from a single male during their mating flight. The great advantage of having such a brief and boring sex life is that the queens and males need only leave the safety of the nest once to hook up. For the rest of her life, the queen can remain securely tucked away deep down in her colony.

DANGLING DESERT REFRIGERATORS

For humankind the sperm bank is a fairly new invention, whereas ants have used their technique for countless millennia. Other modern technologies were also presaged in ant evolution, likewise based on extreme specialisation and the division of labour. A marvellous example is found among honeypot ants, which have mastered the art of long-term food storage without electricity or ice blocks. And masters they must be, since they live in deserts and other places where the food supply is unpredictable. Desert habitats can stay arid for years on end, so their inhabitants must be ready to act quickly when the rains come. The moisture makes dormant seeds sprout and grow into blooming plants. This throws desert-dwellers into overdrive as they rush to gather and store as much of the bountiful food has possible.

Honeypot ants are perfectly adapted to make the most of this situation. Certain workers in the honeypot colony are professional gluttons, capable not only of imbibing but also of storing incredible amounts of food. When the food supply peaks, these individuals are force-fed with nectar, causing their elastic abdomen to swell to the size of a grape. Once they reach full capacity, the sated ants simply hang motionless from the ceiling of a nest chamber, never leaving the colony. Later, when food becomes scarce, they regurgitate the stored nectar and feed their nestmates, enabling them to survive until times of plenty return. Collectively the ant gluttons serve as a living food-storage facility, a communal stomach that can be filled and emptied as needed – the ants' ancient equivalent to modern refrigerators.

CLAD IN PRACTICAL STRIPES

The honeypot solution is so extreme because such species live under extreme conditions. Various common Northern European species are also able to gather food in fairly large amounts, albeit more discreetly, as we can see every summer in our gardens. Stop for a moment by a tree trunk with ants on it. Look closely and you will see that some of the ants going down are slightly rounder than those going up. Ants climb to gather honeydew and nectar, which they temporarily store in their forestomach or 'crop' and later feed to their nestmates, foraging not only for themselves but also for the ants doing other jobs back in the colony. To reduce the number of climbs, the ants eat as much as possible, causing an interesting colour change. On their way up the tree the ants are black, but on the way down they are striped. When their crop is full the exoskeletal plates on their abdomen are pushed apart, making engorged ants appear lighter in colour where the thin skin is visible between the plates. In the ant world, as in our human world, stripes really are extremely practical.

SUICIDE BOMBERS

Human life today revolves around many other features than efficiency. Would such extreme specialisation be beneficial for humans? It might simply create bottlenecks and boredom, given that our production processes and cultural activities go far beyond mere physical survival. Still, there are numerous examples of how toning down our personal interests to benefit the greater good has made a decisive difference, creating situations (for better and for worse) that would not otherwise be possible. Let us take suicide bombers as a case in point; another discipline not reserved for humans. The ant world has martyrs too.

Once again, the self-sacrificing nature of these ants is not just a cultural or functional role but a physical manifestation. In some species the soldier caste can transform themselves into mobile bombs. A special gland rooted in their head can produce certain chemicals –

phenols, terpenoids and aliphatic compounds – which, if bottled, would merit a skull-and-crossbones symbol, marking them 'extremely hazardous'. If the ant is attacked by enemies, these chemicals mix within the gland, causing super-high temperatures that trigger an explosion and spread the toxic cocktail far and wide. The hot, sticky, agonising concoction will kill most of the ant's enemies. It uses this deadly trick to protect its colony and queen. Strategically, when an ant gives up its life, it will take down more than one enemy as its body explodes.

NO TIME FOR CHIT-CHAT

There are obvious advantages to living in social groups, cooperating and dividing labour. But such societies must be extremely well-organised, and one thing is paramount: efficient communication. Humans and ants are expert communicators, but they use very different methods. Humans mainly communicate vocally, with sounds – we speak to each other – whereas ants rarely use sound-based techniques. Rather, they use chemical signals that are captured and decoded by special olfactory receptors found on their extremely sensitive antennae. Ants have a wide variety of glands in their bodies, and they produce chemical scents in various versions and combinations, depending on what they want to communicate. Take alarm pheromones, which are gaseous and thus spread quickly through the air. If a colony is attacked, the front-line ants release these substances to call in reinforcements, and legions of workers surge from the nest to help out on the surface.

An ant can mark a trail of other, less volatile chemical compounds on the ground. Such trails can lead its nestmates to new-found sources of food. Pheromones are invisible, but their presence becomes clear if you find a well-trodden ant path and pull your finger across it at a right angle. Ants arriving at the gap will immediately stop, confused at finding their scent trail gone. It can take several minutes before they once again work their way across the pheromone gap and fix the trail.

It is easy to observe the making of an ant trail. Just offer the ants on your terrace a drop of sugar water or a morsel of moist cat food. The ants that stumble across the new meal will take a bite, then return to the nest dragging their abdomen across the flagstones. At the tip of their abdomen is the exit duct from the gland that produces trail pheromone, allowing ants from the nest to find their way back to the food. Soon a large group of worker ants will gather at the food source, eating from it and defending it against intruders.

Ant collaboration has no recourse to email, telephones or Skype – but ants don't have any use for these things. Scented compounds cover all their needs as they build and operate extensive societies. Their requirement for high-level communication is fulfilled by systems very unlike those in human societies, but nonetheless impeccably efficient.

NATURAL MEDICINE

Social living has drawbacks as well. Many diseases thrive in crowded, close-knit societies. Ant colonies can have up to 20 million members, making it easy for disease spores to move amongst the inhabitants. For a contagious bacterium or virus, a bustling community living in close quarters is paradise. Just think of how successfully a common cold can spread in a crèche or primary school. In our human world the pharmaceutical industry has waged war against certain microorganisms, which in turn respond by developing resistant strains. So far humanity is winning, although new strains of multiresistant bacteria pose no small threat.

Ants also have weapons to do battle at a microbiological level. This explains what could seem like peculiar behaviour. Wood ants collect large amounts of dried resin, depositing small fragments of the stuff in their anthills. Some nests contain as much as 20 kilograms of resin. What they are doing is exploiting the antibiotic properties of resin, the trees' own safeguard against infection, exuded where bark has been damaged or branches broken. Resin also counteracts the bacteria and fungi occurring in anthills. Each time an ant walks over a piece of resin, it comes into contact with the resin's antibiotic substances, which kill infectious spores it may be carrying. Experiments conducted on nests infected with microorganisms show that survival rates in ant colonies with access to resin were twice as high as in nests without access to resin.

Communicating efficiently and preventing disease are two essential issues ants must address, but living in large communities also poses challenges that are far more brutal. This will become clear as we delve into the harrowing topics of ant-scale freeriding, social fraud, state coups and slavery.

FREELOADERS AND FRAUDSTERS

THE COST OF SUCCESS

Understandably, lottery winners often prefer to remain anonymous and keep their good fortune a secret. Hit the jackpot and you will soon find yourself besieged with bloodsuckers and beggars. So too in nature, where this phenomenon is so prevalent that biologists have a specific term for the organisms that practice it: social parasites. Ants are tempting victims for organisms looking for a good host. Reaping the benefits of successful socialisation, ants are now among the planet's largest species in terms of biomass. This gives an enormous advantage to anyone, or anything, that can rob even the thinnest sliver of their sizeable pie.

A wide variety of insects and other tiny creatures – known as 'myrmecophiles' (ant-lovers) because of their special preference for ants – are capable of sneaking into anthills and sponging on their resources. These social parasites strike at the ants where they are most vulnerable. Like hackers, they subtly infiltrate the colony's chemical communications system. Ants use scent-based signals to tell the difference between friends and foes, and only individuals that give off the appropriate and very particular scent of the nest are allowed to enter.

Conversely, any creature that can crack the scent code will be treated like one of the family, finding an open door to the colony's cornucopia of amenities, ranging from shelter and heat to an all-you-can-eat buffet of soft, nutritious ant larvae, ready for the taking.

THE KILLER BABY

In its larval stage, the butterfly species alcon blue is one of nature's most ingenious social parasites. Soon after hatching in the blossom of an herb called marsh gentian, it drops to the ground, where it mimics the scent of an ant larva. The next passing ant – perceiving it as a poor little ant grub, lost and far from home – will pick up the larva and carry it down into the colony. Nestled among the ant brood, the impostor can begin to feast on its adopted siblings. The nurse ants, unaware of the grisly scam, are so taken in by the larva's scent that they even feed it regular ant-food, too. The butterfly-to-be can lie back and enjoy the rest of its larval stage without giving another thought to protection, or meals.

Without dwelling on clever parallels in human society, suffice it to say that fraudsters are not uncommon in the domains of art, research and high finance ... so put on your best disguise and let the fun begin.

THE ENEMY WITHIN

Parasites aren't always total strangers. Oftentimes, in societies formican and human, the most sinister enemies are found within one's own ranks. Certain ants specialise in conquering the colonies of other species, for obvious reasons: An ant colony is a vast, resource-rich community that has invested loads of time and energy in living quarters, infrastructure and vital knowledge. Think about the leafy nests of the weaver ant, or the great mounds of the wood ant. Now imagine how long it takes to clear paths that are hundreds of metres long around or between nests, and think of the huge value of knowing about nearby food sources like nectar-rich plants or aphid colonies. Not to mention the colony's huge labour force, which already lives there.

If you can take over such a wealthy kingdom with little effort on your own part, you've got it made. Besides instantly ensuring your own bright future, you avoid putting yourself in harm's way, as you would do when establishing a colony from scratch. Far less than 1% of the thousands of virgin queens that can set off from a colony on their mating flight each year will reach the egg-laying stage. And even for these fortunate few, the road to a thriving, long-term viable ant colony is long and fraught with hardship.

Initial establishment is, in itself, a complicated process. Usually a young, successfully mated queen finds a suitable location where, on her own, she digs a small hole and barricades the entrance, then lays her first eggs. These she raises through the larval and pupal stages using the nutritional reserves in her body, including the wing muscles she no longer needs. Then, she waits. The miniature colony cannot open until the first worker ants emerge from their pupae. Once they do, they crawl out and look for food to sustain the queen and her new brood. She now has a tiny functional community, but there is still a long way to go, and the colony must invest heavily before reaching any noticeable size. Obviously, any ant that can avoid this perilous part of the colony life cycle will save itself quite a bit of trouble. But how to go about this?

THE STATE COUP

The cleverest ant species simply avoid the start-up phase altogether. Once mated, their young queens fly directly to an existing colony, hoping to overthrow the regent and take over the entire colony in one fell swoop. First, the aspiring queen must make it past the gatekeepers, which will check to verify her scent. We still don't know precisely how species with 'pirate queens' do this, but we do know they dupe the guards. In one such species the attacking queen seizes a worker from the colony and bullies it a bit before setting it loose. The poor worker races back to the nest to report the incident, and as the rumour of an external attacker spreads, its nestmates go wild. Amidst the confusion of worker ants racing about to find out what is going on, the invading queen slips unnoticed into the heart of the colony.

The next step in any *coup d'état* is getting rid of the incumbent ruler. Here, too, different pirate species have different strategies. One of the more subtle methods is based on a sort of lobbyism where the invading queen persuades small groups of the host colony's workers to change sides. Again, the exact details of the strategy remain unknown, but basically she passes her antennae across a small group of workers and entreats them to feed her. This evidently buys their loyalty. Gradually the intruder wins over more and more worker ants, which begin to work for her while ignoring the reigning queen – their own mother. Ultimately they not only ignore their old queen but kill her. The coup has succeeded, and the new queen never even got her hands dirty.

Not so with the 'headcutter ant', a species whose invading queen will cut to the chase, dashing straight down to deal with the incumbent ruler herself. Before the workers can overwhelm her, she climbs up onto the reigning queen's back, where she is safe from further attacks. From this comfortable perch she spends the next couple of days slowly nibbling her way through her rival's neck – so if not dirty hands, she certainly ends up with dirty mandibles. Rather bizarrely, a few more days must pass before the beheaded sovereign has been truly ousted. Despite this book's unfailing enthusiasm for ants, I must admit that each individual has very few brain cells. So few, in fact, that an ant has no problem living on without its head, at least for a time. Unfortunately, ants need their head to take nourishment ... so the proclamation after this sort of ant coup is: The Queen is dead (of starvation). Long live the Queen!

SLAVERY

Another modus operandi used by social parasites is slavery, a behaviour likewise not confined to the human race, although that is small consolation to us. Yet interestingly there is scant evidence of racism among ants, which subjugate other species as willingly as their own. Slavemaker ants are only interested in a colony's brood. Unlike adult workers, the immature larvae and pupae have a neutral scent that is not colony-specific, so they can be imported into a new nest without being perceived as enemies.

When the abducted pupae open, the emerging ants take on the scent of the foreign colony and blend in with the rest of the community, replenishing the labour force for free. By contrast, captured larvae face a dismal fate. In slave raids, only pupated individuals are allowed to live. Grubs need feeding, which costs the slavemakers time and effort, so their captors eat them instead.

One slavemaking ant species is *Formica subintegra* – a cousin of the European red wood ant – which uses chemical warfare when raiding other colonies. Much like the suicide-bomber ants, this species has also developed a greatly enlarged special-purpose gland in its body. The chemical compound from this gland is released during an attack, and as if by magic the ants defending the colony panic and spread out above ground instead of running down to defend their brood. This makes it much easier for the slavemakers to reach the brood chambers and steal pupae. Some species have even become so dependent on slaves that they cannot complete their life cycle without them.

Denmark, like much of the Northern hemisphere, also has ant species that raid other colonies, and since it is easy to keep ants in a terrarium, there is virtually no end to the interesting behavioural experiments you can do right in your own living room. Yes, this is your friendly neighbourhood scientist saying: *Do* try this at home! These geographical areas also have species that use queen-killing as a propagation strategy. Practices of slavemaking and state coups both lead to colonies that consist of two ant species living in the same nest. So next time you move a paving stone in your garden and find a nest of black *and* yellow ants, you will know it is not love that brought them together.

BE FRUITFUL AND MULTIPLY

Through the evolutionary development of ant communities, much of the world's biomass has become social. Admittedly, societies face challenges large and small, including savvy systemic parasites. Nevertheless, this does not alter the fact that social organisation is a fundamental part of what makes ants and humans so successful. Other factors also play a role, the most important being an impressive, and shared, ability to manage energy resources, understood as heat and food. This has enabled both races to conquer some of the trickiest locations on Earth, where such capabilities are absolutely crucial. To maximise biomass, you must be able to live in as many different habitats as possible, and for any race seeking to proliferate and fill the Earth – consciously or instinctively – diversification is a must.

COLONISING THE CONTINENTS

THE CONTINENTAL CHALLENGE

Humankind and antkind have conquered every continent on the globe, except Antarctica. Thanks to our adaptability we can survive in deserts and swamps, on mountain tops and valley floors. If we study the highly varied physical conditions in different geographical areas, one factor is especially problematic: temperature. If you want to meet the continental challenge, you must be able to cope with temperatures ranging from –89°C to +58°C, the records measured in Antarctica and in Libya. This calls for enormous adaptability. Humans have what it takes, and so do ants, although our coping strategies are very different indeed.

HEAT, HEAT AND MORE HEAT

Early humans were first able to move into colder regions when clothing was invented. Throw on a few furry pelts and you're set to go a few degrees further north. Around the same time, humans made what is probably our greatest discovery ever. We found out we could control fire, allowing us to conquer the last few degrees of latitude towards the poles. However, even without fur coats and fireplaces, humans are quite well-equipped to survive.

Unlike ants, human beings are endothermal (formerly called 'warm-blooded') creatures. We are blessed with a constant body temperature that enables us to keep our enzyme systems and activity levels up and running, even in ambient temperatures far below the tolerance threshold of our enzymes. We obtain the energy to fuel this system from the food we eat, so theoretically we can eat our way through a cold spell. Hardly surprising, then, that seal blubber is a favourite dish among the Inuit peoples of the Arctic.

TWO WAYS TO KEEP WARM

But for ants, low temperatures are a challenge. They are ectothermal (or 'cold-blooded') and automatically take on the temperature of their surroundings. In the cold, ant enzymes stop working, bringing them to a physical standstill and preventing them from moving, no matter how hard they try. But ants don't give up so easily. In severe weather they can enter a dormant state and sleep their way through a cold spell. What is more, they can raise the temperature in their surroundings to make warm spells last as long as possible. Welcome to thermal engineering, ant-style.

Consider the mounds of the wood ant. They are actually little heating plants made of pine needles, twigs and other organic material gathered into a heap. As amateur gardeners know, when such material is piled together the result is a compost heap, which becomes warm inside. Real compost enthusiasts can even generate enough heat to 'cook' undesirable seeds and weeds to death.

Similarly, ants use biofuel to generate energy in the form of heat. Unlike human biogas plants, an anthill's combustion takes place in the presence of oxygen, which is why the biomass in an anthill is converted directly into heat instead of gas. Nevertheless, in both cases we are looking at a bioenergy process that, in cold conditions, keeps anthills and nests significantly warmer than the air outside.

A MOUNTAINOUS ANTHILL

Not all ants need anthills. Far from it. Anthills rarely occur in the tropics, but they sometimes grow to colossal size in the northernmost reaches of the wood ant's domain. In Norway, some species build anthills that stand taller than an adult human – which, from an ant perspective, is positively mountainous. The Norwegians even have a competition each year to locate the tallest anthill. The record, observed in the Trysil area, is held by an industrious ant colony whose mound has a circumference of 11.2 metres (36.75 feet) and a maximum height of 2.95 metres (9.67 feet) above ground level. If you have ever observed a wood ant lugging a tiny twig around the many obstacles on the forest floor, you will appreciate that even modest anthills are the result of great investments, in terms of the time and energy spent to build them. Even so, their function is far more valuable than the investment itself. The heat within the mound enables its inhabitants to survive and thrive, even after the onset of cold weather.

SOLAR-PANEL SYSTEMS

Compost heating is not the only amenity an anthill can provide. Its inhabitants can also raise the temperature by exploiting energy from the Sun. By raising its hill above ground level, an ant colony at northern latitudes, say in Denmark, will have the upper surface of its nest catch far more incoming sunrays than a corresponding circle on the flat ground. In the dawn hours it can absorb up to three times as much heat. Even in the midday hours, with the Sun near its zenith, an anthill will gather 20% more energy than a flat, ground-level alternative. But the most important heat radiation arrives in the early hours of the morning, when temperatures are lowest in the 24-hour diurnal cycle and the extra heat is needed most. The urban plan of an ant city meticulously lays out its landscaping to address this need. If you examine your local colonies of wood ants, you will find that virtually all are located in a clearing or near a path that lets sunlight enter, and that most are positioned to catch the morning rays. Later in the day, if the Sun starts to bake down, the nest can become quite hot. If it gets too hot the ants simply move to the shady side of the mound or turn on the anthill's ventilation system by opening its many auxiliary entrance holes.

An anthill's function as a solar panel is also clearly seen in the almost ubiquitous colonies of black garden ants, which don't build hills atop their subterranean nests if they live beneath a well-cropped lawn. But try letting the grass grow up around an anthill, which will prevent the sunrays from warming the nest. You will soon see the tiny excavators begin their earth-moving work, building a cone or dome from loose particles of soil supported by the blades of grass, all in the interest of capturing precious energy from the Sun.

PAVEMENT ENEMY NUMBER ONE

The paving and flagstones we lay in our gardens are unaffected by most external influences, but one force of nature can actually get the better of them: ants. As a result, many of us are prepared to douse such outdoor living spaces with generous amounts of poison. How frustrating they are, these persistent little pests! Meddling with our landscaping and leaving their messy piles of sand on our nice, smooth patios, where we have every right to sit, unperturbed, and enjoy the sights and sounds of Nature tamed and trimmed.

But why are ants so fond of flagstones? Once again, because they have conquered continents where heat is in short supply. In their quest to keep warm they have discovered that paved areas act as ready-made solar panels, complete with a stable roof. Such surfaces are freely exposed to sunrays. In the daytime they absorb heat, retaining it for many hours and allowing the nest below to remain comfortably warm far into the night. If a paved surface has the added advantage of sitting directly against a foundation – whose vertical wall, like an anthill, can catch the early-morning sunlight – it is hard for an ant to find a better nest location. So, when you find the ants have been fiddling with your flagstones, they're not being malicious. They're just cold.

The ants' partiality to paving stones is notorious. So much so that my mentor and faithful guide into the fascinating world of ant (and university) communities – Mogens Gissel Nielsen, the grand old man of Danish myrmecology – has spent decades studying the paved surfaces around the university's nature-lab centre in hilly Mols, part of the Djursland peninsula. In fact, he has peppered the hillsides in the centre's vicinity with paving stones, knowing that as soon as he lays one down, the ants will arrive. In this way he has turned his annual ant-spotting routine into a myrmecophile's walk in the park – or rather, the hills – as he tracks the prevalence and shifting of species in the area. He simply wanders around peeking under all the paving stones.

CENTRAL HEATING

I have already described how ants can produce and gather energy. Fascinatingly, they can also move energy around, almost as we humans do in central-heating systems. If you go for a walk past your favourite anthill on a cold but sunny spring day, you will see large patches of worker wood ants in the sunny spots atop their mound. Here they stand, close together, warming up their black abdomens, which, just like our human bodies, mainly consist of water. These ants are acting as live hot-water bottles, storing heat in their bodily fluids. Temperatures drop quite slowly in moist substances, giving the ants time to crawl deep down into the nest. There they give off their heat to the queen and the brood, which lie snugly tucked away in the darkness, protected against enemies, in a humid atmosphere that can keep a larva's delicate skin from drying out. This way the worker ants effectively serve as a radiator system, heating water in one location, then transporting it to the chambers that need warmth.

ANTS ON HOLIDAY

Biofuel, solar panels, hot-water bottles and radiators? Yes, the mechanisms ants employ to handle their heating needs are numerous and elaborate. When ants enter our homes, they largely want the same things we do, and problem incursions into human dwellings mainly occur when the weather starts to turn cold, but not cold enough for the ants to go dormant. In the spring and autumn months they seek out heat indoors, attempting to prolong their season – much like humans who live in chilly northern climates fly south to enjoy warmer weather.

The general ability of ants to manage heat has given them access to a vast geographical area, increasing their biomass immensely. The final evolutionary step towards winning the biomass championship also has to do with finding and controlling energy, but in a different manifestation: food.

ANTS IN THE FIELD

IT'S AN INDIGESTIBLE WORLD

Take a look around. Unless you happen to be sitting in an urban jungle or sweating your way across a desert, you will see that the world is mostly green. There are plants everywhere. One key to success is gaining access to the boundless resources plants represent. If you could eat plants from root to tip, you would never go hungry. Instead of scurrying around in search of your next meal you could spend your time procreating, thereby increasing the biomass of your species. Plants are all around you, so there must be a hitch, an obstacle that prevents you from doing just that.

As it happens, there are several obstacles. First, plants protect themselves with certain tissues that produce toxic or antidigestive chemical compounds. Second, plants have a big advantage because most of their dry matter consists of a combination of cellulose and lignin, known as 'woody matter', which very few organisms can digest. Woody matter is built from the sugar molecules plants produce through photosynthesis. Regrettably for would-be diners, these sugar molecules are chemically linked into long chains, making the woody matter hard and tough, and giving the plants strength and support. This is a smart strategy if you want to grow into, say, a tree trunk. Not many organisms can make enzymes capable of cutting these molecular chains and converting them back into sugars, so woody matter is largely indigestible. This means just a small proportion of plant-based biomass is readily accessible as food. Nevertheless, it is possible to gain access to the plants' resources if one develops agriculture – yet another sphere of activity where ants and humans are among the most skilled creatures on the planet.

JOINING FORCES WITH MICROORGANISMS

Many human agricultural practices are based on alliances with fungi and bacteria, some of the few organisms that can actually break down woody matter. For us, the road to riches goes through cattle. Bovine stomachs host microorganisms that allow us to access the energy bound up in blades of grass. This road to riches has a toll house at every crossing, however, and the microorganisms demand their fair share of the energy. Still, from the cattle's point of view there is a profit to be made. They stay healthy and contented on this diet – and when we drink a glass of milk or eat a steak, we are essentially enjoying reworked grass.

This process of converting grass into beef is costly, since the microorganisms generate many other substances besides sugar when breaking down the woody matter. Much of the energy from the grass is bound in methane, a gas that makes our extensive farming of cattle significantly impact the environment. Cattle get rid of methane by farting (which, in the bovine anatomy, is actually burping).

And because methane is a highly efficient greenhouse gas, these biological emissions put pressure on the climate. However, despite the energy it costs, the strategy seems to work, given that humans and cattle are among the top four on the current biomass list. Indirect access to plant resources, as described above, is probably part of the explanation for this. What is more, the huge potential of plant energy is illustrated by the fact that those of us who drive cars with combustion engines are actually filling up our tanks with plants, now in handy, liquefied form after spending several million years underground.

LARGE-SCALE MUSHROOM FARMING

Some ant species also access plant resources by means of other organisms. Their method is fungus-farming in underground chambers. If you have ever visited a tropical biotope park with an indoor leafcutter colony under glass, you may know part of their remarkable story already. This ant species invented agriculture some 50 million years ago, around the time certain mammals began to develop into early primates. In other words, ants were farming millions of years before any faintly humanoid life form ever walked the Earth. To make a long story short, leafcutter ants have an impressive track record in farming, and they have not wasted their time. The system they have developed to cultivate their particular, preferred fungus is not just unique. It is perfectly adapted and highly intensive. Judging from the name of the species you would think these ants ate leaves, which they don't. The leaves they cut are carried home and used for compost. First, the indigestible leaf matter is chewed into a pulp, which is then seeded with mycelia (microscopic thread-like structures that spread underground) from the fungus that provides leafcutter ants with the only food they ever eat. As the fungal mycelium network grows, it forms an extensive system of threads, the ends of which swell and grow – and *voilà*: Dinner is served. The fungus has broken down the plant's defences, so the ants can harvest fungus meals in the comfort of their own nest. The leafcutter ant's farming technique – unmatched by any other species – is its ticket to a world of plenty, and its strategy for avoiding the many toxins used by plants to protect themselves.

METAL JAWS

This cultivation process is organised much like an industrial assembly line. Each ant caste has its own task. Medium-sized ants shear the leaves off the trees using their mandibles and massive jaw muscles. Smaller individuals cut the leaves into manageable pieces. Slightly smaller ants chew these pieces into compost pulp. The smallest inhabitants in the colony distribute the fresh pulp around the fungus garden, seed it with mycelial filaments, then tend and harvest the fungi. They spend their entire lives deep underground, never leaving the nest. To create this intricate process, leafcutter ants have had to develop a whole range of worker castes in different sizes. As a result, the colony's smallest ants are dwarfed in size by even the head of the largest workers, which weigh 300 times more than their diminutive sisters. The ants that spend all day cutting plant material even have metal compounds built into their jaws, which consist of 16% zinc. The zinc makes their mandibles harder and sharper. These effective shearing tools are so specialised they limit the workers' function in life to a single, trivial cutting task.

AGRICULTURAL CHEMICALS

Leafcutter ants now eat only this one fungus, and the fungus cannot survive without being tended by the ants. Just as human farmers must limit weeds in their crops to optimise the harvest, ants must keep their fungus gardens and compost free of competing fungi. Weed-free fields and flowerbeds are always a challenge, as farmers and gardeners well know. Chemistry can help keep weeds at bay, as it does in much modern-day farming above ground.

Although leafcutter ants use their mandibles to mechanically remove foreign mycelium, they also use chemistry as an aid. Endowed with a special gland, they can release three different compounds targeted at fungus-tending. The gland produces the chemicals phenylacetic acid and myrmicacin, and the plant hormone indoleacetic acid. Respectively, these suppress bacterial infections on the fungus, keep foreign fungal spores from germinating, and stimulate fungal growth. In effect, the ants tend their subterranean gardens using antibiotics, fungicides and growth hormones – the same cultivation strategies humans use in intensive conventional farming, but infinitely older.

AGRICULTURE RULES

Picture the wide, green expanses of grain fields that cover Denmark or some other agricultural countryside you know. Then call up a mental image of an uncultivated meadow (a rare sight in many modern landscapes) where grasses, herbs and other plants grow wild. A comparison makes it clear that when an organism systematically begins to cultivate the crops it can best digest, its food supply increases dramatically, in volume and in dependability. As a food source, a natural meadow cannot compete with a field of efficiently farmed grain. This comes at the cost of less natural diversity, but agricultural methods have given humans a huge advantage in conquering the world.

Around 12,000 years ago, after humans discovered agriculture, our population exploded and has since followed an exponential growth trajectory. This is because we have continued to optimise production (greatly aided by fossil fuels). Agricultural practices enable us to increase plant production, and by collaborating with microorganisms and their animal hosts we can go one better and additionally gain access to indigestible plant components. In this light it is probably no coincidence that, as farmers, leafcutters rank among the ant species with the largest communities.

ANOTHER CLEVER TRICK

Leafcutter ants are not the only insects that have cracked the plant kingdom's security code. Another clever way to access a plant's well-protected resources is to capture its sugar before the plant converts it into woody matter. As mentioned, plants create sugar through photosynthesis, which takes place in their green leaves. The sugar is then transported through the plant's veins to the tissues that need the sugar. These veins lie sheltered beneath the plant's bark, but certain animals have specialised in reaching the plant's pockets of sweet juices. Certain insects with beaked mouthparts (an order of insects that includes aphids, scale insects and cicadas) have evolved mandibles that are shaped like a stiletto and can penetrate several centimetres of tough plant matter. In this fashion they can locate the sap in the plant's veins and siphon off nutritious food.

In this food-hunting discipline ants are out of the competition from the start. Their biting, cutting mouthparts are useless for drilling deep into plants, but they have entered the race as outsiders, thanks to another clever trick. Once again they exploit their potential for creating alliances with organisms that can do what they cannot do themselves. Many ant species live in symbiotic relationships with beak-mouthed insects, 'milking' their partners and removing the excess sugar we know as honeydew.

ANIMAL HUSBANDRY

Aphids provide a good supply of sugar. When an ant strokes an aphid's back with its antennae, the aphid reacts to the cue by exuding its honeydew. The ant carefully sucks up the droplet, avoiding waste and storing the sweet fluid in its crop. The ants, eager for more sugar, optimise living conditions for the aphids, helping the aphid colony grow. One way ants do this is by protecting the aphids from enemy attacks. A hoverfly maggot attempting to eat the aphids will immediately feel the fury of any ants nearby. This is incredibly helpful for aphids, which have partially lost their ability to flee attackers, being effectively stuck in place with their mouthparts buried deep in the plant's tough surface. It can take them several minutes to extract their sucking apparatus, so aggressive ant guardians are valuable companions for a hard-working aphid. This combination of ant protection and aphid equipment makes for a fully symbiotic lifestyle that enables both species to access a well-protected plant-based bonanza.

Aphids, being built to sit in one spot and suck sap, are not very mobile, so the ants move them when the need arises. If an aphid colony grows so large it risks overburdening its resident shoot, the ants pluck some of the aphids off the shoot and tote them off to fresh, unexploited shoots, just like herdsmen moving cattle or sheep to new pastures. And the comparison does not stop there.

Like humans, ants are not always trustworthy partners. When the aphid population reaches a size that matches the ant colony's needs, the ants no longer have to spend time tending them. At this point the ants could simply choose to stop their animal husbandry, but they take the symbiotic relationship further. They begin to eat from their aphid herd, thereby also accessing the plant proteins, the aphids have spent their entire lives diligently filtering out of the sap. So just like we choose whether our cattle herds will supply us with milk or beef, the ants choose whether their herds will supply them with honeydew or aphid steaks.

THE QUEEN'S DOWRY

In some cases the symbiotic relationship between ant and beak-mouth has grown so intimate that virgin queens from the ant colony are given an aphid or louse to take along on their mating flight as a sort of dowry. The queen caste has developed a tiny abdominal pocket that accommodates the beak-mouth during flight. This way, when a mated queen settles down to establish a new ant community, she has a little nest egg: access to honeydew from her pet aphid – which can make a crucial difference to her success.

Through evolution, some species of beak-mouthed insects that live as symbionts with ants have completely lost the ability to protect themselves against enemies, nor can they rid themselves of their honeydew. They have become so dependent on their ant symbionts that they cannot biologically function without them – so the ants have bred species of insects that can only live in their company. Their beak-mouths are, in essence, an example of animal domestication, a phenomenon observed in just one other race: humans.

With so many similarities, one would think humankind felt a certain kinship with ants. This is by no means the case, but prejudices aside, let us now reflect on the broader question of potential interspecies cooperation.

FROM FOE TO FRIEND

UNCLE PREBEN'S WAR

My energetic, persistent Uncle Preben has never let anything stand between him and happiness. One year, happiness meant taking back control of his greenhouse. The ants had invaded in force, and to this day my uncle claims they had actually started moving the stepping stones he had laid out among his prize tomato plants. If so, they must have been Hercules ants – a species not found in his part of Denmark. At any rate, no respectable human should have to stand for such antics, and because that blasted ant poison at the hardware shop cost a fortune and didn't work anyway, my uncle saw no other way but to take matters into his own hands.

Having been a driver all his life, Uncle Preben knew a thing or two about combustible fuel. He marched off to the shed, fetched his spare-fuel canister and poured its high-octane contents directly over the beastly insects. Confounded by the fact that they did not just curl up and die, he decided they deserved to go up in flames. Fishing his lighter out of his pocket, he prepared to burn the ant colony like a witch at the stake. But Uncle Preben had forgotten an important detail about liquid fuel – and his greenhouse exploded.

He returned from the hospital to find that the ants had built a small anthill to keep warm in their now generously ventilated greenhouse.

WANT TO MAKE FRIENDS?

Uncle Preben's perception is typical of the human–ant relationship. Everybody hates ants. They are our enemies, not our friends. If you google the word 'ants', all the ads that pop up will be for companies that kill ants, one way or another. None of these companies sell ants. However, as long as we continue to regard these animals as our enemies, we are pitting ourselves against a formidable foe, battling a biomass larger than our own, and criticising a creature that possesses many of the characteristics that benefit humans most. We face a tough match against a tenacious opponent, and we will never be able to get rid of ants completely. We can reduce the problems they cause, but the moment we turn our backs on them, we're back to square one.

But perhaps we can find inspiration in the martial arts, which urge practitioners to exploit their opponent's own strength by turning it against them. We can apply a similar philosophy to our dealings with ants, identifying their strengths and exploiting them by directing ant-power against other problems we have in the human world. That way we can recruit ants as partners. Suddenly, rather than an invincible opponent we will have a strong, stalwart ally.

WEAVER ANTS AS PEST CONTROL

Years of research into the life of weaver ants reveal that they can become valued partners, although at first glance they do seem like a nuisance. Weaver ants look for food close to their nest, up in the trees, where thousands bustle about in the treetops. This is common knowledge among tropical plantation farmers from Asia to Africa. When they begin to pick the ripe fruit, the weaver ants immediately attack and defend their territory. This is painful for human fruit-pickers, especially if one biting ant is suddenly joined by 200 nestmates. All other things being equal, it is easier to harvest fruit from trees with no ants, so most tropical farmers regard the weaver ant as a notorious pest that is best avoided.

However, all other things are *not* equal. The point is, trees without ants bear far less fruit. The problem seen from a human angle – that the ants are everywhere, and that they bite – is a great advantage for the tree. You see, the busy ants don't just bite farmers and fruit pickers. They also bite all the pests that impede good pollination, fruit formation and crop size. Seen from that angle, suddenly a host of busy, biting ants is an advantage.

Studies show that ants can combat 50 different types of pests in at least 12 different types of common plantation trees. This can benefit crops such as mango, cashew, coconut, cocoa, coffee and mahogany. Using ants as a biological pesticide can even prove more effective than the conventional chemical poisons so widely used today. The world's leading authority in this research field, the Chinese entomologist Dr Renkang Peng, has demonstrated that cashew and mango production in northern Australia can be increased, boosting net plantation income by over 70% by simply installing weaver ants in the trees and abandoning chemical pesticides. Ants are far more efficient and much cheaper than chemical alternatives. What is more, because weaver ants are indigenous to the tropical regions of Australia, Asia and Africa, farmers in these parts have easy access to a naturally occurring animal partner with a huge potential to improve crop protection.

FROM DOUBLE TROUBLE TO DOUBLE UP

The ants are a double asset for plantation owners. Being edible, they also provide a rich source of protein. Thailand has a long tradition of harvesting ant larvae, which fine restaurants serve as a popular and expensive delicacy. And before you start making faces at the thought of eating insects, here's a newsflash: While it is true that typical Western diets do not include insects, this actually makes us a minority, globally speaking. Entomophagy – the practice of eating insects – is widespread in virtually all other cultural spheres. The renowned and highly innovative Nordic-food restaurant Noma, located in Copenhagen, actually made headlines after putting a dish on the menu featuring live, naturally orange-flavoured ants. However, considering that these ants conceivably qualify as the most expensive ingredient in the world (with the possible exceptions of connoisseur saffron and

24-carat gold leaf), they will hardly become a new Danish speciality any time soon.

I had the privilege of conducting a research project to find out whether humans can feasibly harvest ant larvae in a plantation while the ant colonies are being used for plant protection. The answer? Yes! If left alone, the queen larvae customarily harvested in Thailand would leave their birth colony once they had reached maturity. In other words, the harvested individuals would have flown off anyway, so harvesting them does not adversely affect the number of worker ants – which are the ones that attack pests in the trees. Our conclusion was that harvesting queen larvae does not interfere with a weaver-ant colony's efficiency in controlling pests.

Hence, the result of our research was positive: It *is* possible to sustainably combine ant harvesting and biological pest control. Here we had set up a system where ants not only removed pests from the crops *human* farmers wanted to harvest. The ants also converted these pests into a protein source that was edible, even desirable, for humans. A problem (the pests) had been transformed into a solution (tasty proteins). The ant-larva yield ranged from 30 to 115 kilograms per hectare of plantation. In terms of efficiency and volume, this is no match for a pig or poultry farm. Then again, the ant larvae are a completely free by-product and potentially a valuable supplement for a subsistence farmer in the tropics.

Our project is currently setting up ant farms in Africa that can produce weaver-ant colonies for local farmers.

What we are doing is domesticating the ants, just as our forefathers once domesticated honeybees and silkworms.

It is inconceivable to me that we should be unable to similarly exploit other ant species, if only we gain enough insight into their biology. If we study the existing ant literature, we find that the vast majority of ant species possess beneficial qualities when it comes to fighting pests. Some seem almost tailor-made for the task. What is it, exactly, that makes them so well-suited for this work? There are many answers to this question, but three characteristics of formican expertise are especially noteworthy: communication, division of labour and cannibalism.

LIFE IN THE RANK AND FILE

Thanks to their social lifestyle, ants have developed the ability to communicate quickly and efficiently. This means they can capture and kill far larger prey than other animals their size. A lone ant that finds a large potential victim begins by gripping the victim and using itself as a living shackle, sinking its mandibles into the prey and anchoring its own legs to the surface beneath it. While this is going on, the ant calls for help by releasing a gaseous pheromone. The scent spreads through the air, reaching nestmates that respond by rushing over to help. Ants can work together to kill large intruders, but they can also dispatch small victims on their own. Unlike other beneficial insects such as ladybirds, which eat only aphids, ants can hunt and combat a wide variety of pests.

The ant colony's division of labour, with just one egg-laying queen, also makes it strong. It is no major problem for an ant society to sacrifice the lives of rank-and-file members. With undaunted zeal they will attack anything that intrudes on their territory. (Remember that picnic you accidentally laid out too close to an anthill?) Even much larger animals that are not potential prey for the ants must bow to their wishes. Certain fruit-loving bat species, for instance, have learned to avoid fruit trees with weaver ants because they know they will be attacked if they try to feed there.

Another advantage is that ants do not shy away from cannibalism. They will eat their own brood if there is a food shortage, helping the colony make it through hard times when other beneficial insects would die of starvation. We have found that ant colonies introduced into plantations rarely need replacing, since they are able to survive when food becomes scarce. This makes ants a cheap, low-maintenance solution since they require no special care and no periodic replacement. Also, and perhaps most importantly, the farmer does not need to monitor his crops to detect pest attacks, because the vigilant ants are always there. Ladybirds, with their restrictive, aphid-based diet, die of starvation once they have solved the problem and the aphids are gone. When the next aphid infestation comes, it's back to the ladybird dealer for a new batch.

ANTS ON THE ALERT

We and the aphids are not the first organisms to discover the power of ants. Certain plants also enjoy their protection. In tropical fruit trees, weaver ants help keep bats and other hungry guests away, and even some plants in northerly climates have developed their own ways of cooperating with ants. In the early spring, take a close look at new leaves on a cherry tree. Where the leaf narrows into the stem you will see a pair of small, red, bud-like protrusions. Often there will be an ant there too. The reason is that these protrusions, called extrafloral nectaries, exude sweet nectar that is irresistible to ants. This is how the tree pays its bodyguards. Without any interference or good advice from researchers, cherry trees have found out that things just run more smoothly when you're on friendly terms with your local ant community.

It works like this: As the ants ramble around in the tree's canopy drinking nectar, they also patrol the entire tree. If they encounter other insects along the way, they either chase them away or kill them for food. Since spring is also a time when many insects are leaf-eaters, this arrangement causes the ants to remove the plant's early enemies. Later, when the leaves are mature, they form their own chemical defences and no longer need protection from the watchful ants. At this point the cherry tree conveniently stops producing extrafloral nectar, so if you want to observe this phenomenon you will have to look while the leaves are fresh and young, although the red protrusions are visible all summer long.

NATURE AS A TEMPLATE

The way plants use ants is an example of biological pest control, utilising natural enemies. Plants have been doing this for millions of years. When humans work with ants, as in an agricultural improvement project, we are not breaking new ground. Rather, we are copying a model that has already developed during the evolution of certain plants. In this case, copying is okay. It is even comforting, for it gives us excellent long-term evidence that ants can serve as a pest control measure in human agriculture. If, during their evolution, plants could 'afford' the cost of developing extra nectar sites and paying the ants with nectar – using resources the plants themselves could have used to grow – we can see it is an enduring strategy. The fact that it evolved must mean that plants using this strategy did better than those that did not use it. Consequently, we can be fairly sure that ants can be used to enhance plant production.

As noted, the challenge lies in identifying the right ant–plant combinations and making systems as efficient as possible. One way to do this is by creating optimum conditions for ant colonies as we grow crops.

INVISIBLE FARMHANDS

Ants can do more than curb pests. In Australia, where much of the land is arid, research has shown that wheat yields increase by 36% if there are ants living in the ground. Their underground tunnels help the soil absorb and retain much more water when rain occasionally falls, so that it benefits the plants instead of simply running off into rivers and oceans. Here we have ants performing the same important function as earthworms, invisibly tilling the soil and helping the plants that grow there.

At the same time the ants apparently fertilise the plants around their nests. As their waste products decompose, important nutrients are released, benefiting nearby vegetation. Epiphytes are a good example. They live on (but not off of) other plants, and they have difficulty finding enough nutrients because they have no roots in the ground, where most of the minerals they need are formed. But they too have partnered with ants to solve their problem. Epiphytes have special hollows that are well-suited for ant waste, and once the ants begin to use one as a rubbish dump, the plant is home free. The epiphyte's trick? The walls of these hollows consist of special tissue through which the plant can absorb most of the nutrients it needs directly from the decomposing waste. In the long term, humans might even be able to put ant droppings to good use in certain agricultural applications.

A FUTURE WITH ANTS

In the early 1980s – back when Greenpeace was expanding their fleet of Zodiacs faster than I could sell their promotional stickers – I read a typical Scandinavian tome from that idealistic era. Appearing in English in 1973 as *Man's Place: Intelligence in Whales, Dolphins and Humans*, this book was the work of the Swedish writers Karl-Erik

Fichtelius and Sverre Sjölander. The message hinted in the title stuck in my impressionable young mind: We humans may not be as smart as we think we are. Whales may be smarter.

Now a full-grown naturalist myself, I would like to dedicate this humble volume to a group of social creatures at the opposite end of the scale: ants. Yet my conclusion is quite a different one, namely that we humans may not be as *unique* as we think we are. Ants also possess many of the capabilities we take pride in, and which make antkind and humankind biologically successful.

In the course of our evolution we have repeated many of the successes pioneered by ants, often extending and refining them. But we can go further. If we can begin to understand ant biology better, we can move beyond where mere inspiration might take us. We can begin to exploit their huge potential.

It is a fair claim that collaborating with nearly one quarter of the Earth's animal biomass has much to offer. Just think of what we might achieve as we gradually learn to harness the capabilities of ten billion, *billion* ants ... As agricultural assistants, ants are incredibly practical and industrious. Thanks to their diminutive size and huge numbers, they can do what we cannot: be everywhere at once, and deal with things too small for us to handle. The possibilities, like ants themselves, are legion.

If we want ants out of our homes, we must study them to find out what makes them tick. If we can identify the chemical compounds of slavemaker ants that make their victims flee in panic, perhaps we can come up with a non-toxic substance that gets rid of unwanted ants. And if we can take what ants covet indoors and move it back outside, we can induce them to move out. Whether we work with ants or against them, learning about their lives is the key to controlling their enormous biomass.

For better and for worse, we are fellow species on this planet and we have much in common. By gaining insight into the lives of ants, humans can learn about what long-term biological success really means – and get useful tips and tricks for our own quest to achieve it.